Farmer Mike

Farmer Mike Grows Giant Pumpkins

by **Mike Valladao**
Illustrated by **Larry Sippel**

QuarterCircleV Publishing
San Jose, CA

Farmer Mike Grows Giant Pumpkins

Copyright © 2008 by Michael Valladao
A Farmer Mike Awareness™ Book

Published by
QuarterCircleV Publishing
3636 Sweigert Road, Suite 100
San Jose, CA 95132
www.QuarterCircleV.com

Printed in China

Library of Congress Control Number: 2008907416
ISBN: 978-0-9820311-0-0

Dedicated to the
coastside community of
Half Moon Bay

Every year, Farmer Mike grows really **big** pumpkins.

How does he grow them so big?

Well, there are many types of pumpkins.

Some are small and only grow as big as an apple!

Some are larger.
And some are much larger!

But the largest pumpkins of all are Atlantic Giants! These are the type Farmer Mike grows.

In late Autumn, Farmer Mike takes seeds from the biggest Atlantic Giant pumpkin he can find.

8

He washes
the seeds,

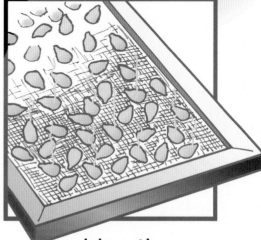

dries them
in the sun,

and puts them
safely away
until spring.

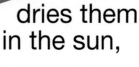

During the Winter Farmer Mike prepares the ground.

For pumpkins to grow big, they need rich soil with lots of organics.

Organics are natural items that come from plants and animals.

Farmer Mike uses a shovel to mix in chopped up leaves, vines, and even some poop from chickens and cows.

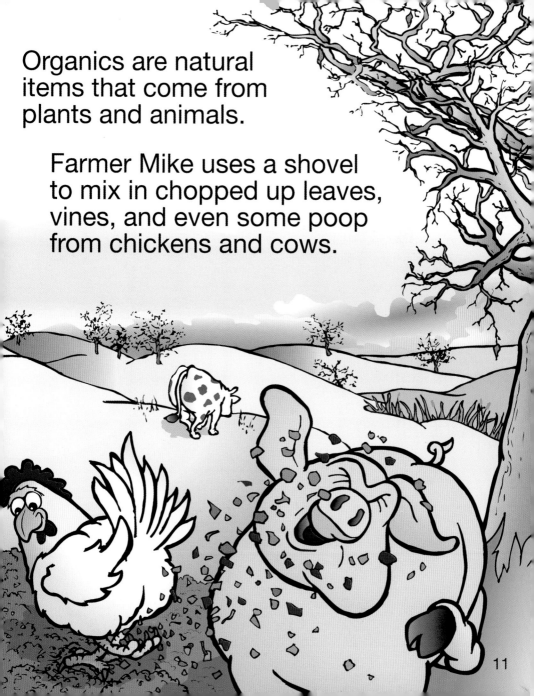

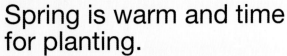

Spring is warm and time for planting.

The root only grows from the pointed end of a pumpkin seed, so Farmer Mike always plants his seeds with the tip down.

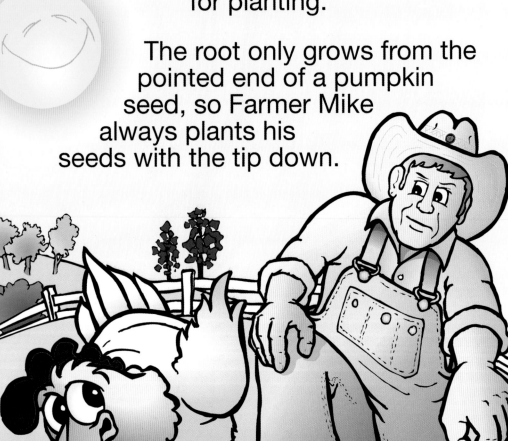

In a few days
the seed begins
to sprout,

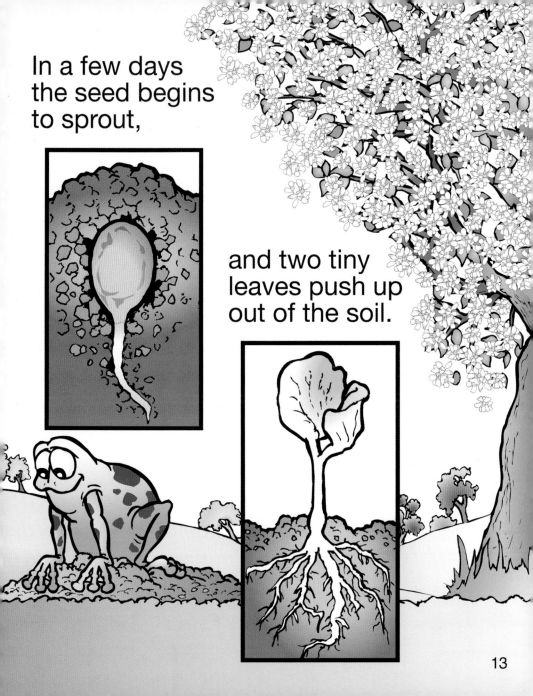

and two tiny
leaves push up
out of the soil.

Like all living things, a pumpkin plant requires food, water and sunshine.

Farmer Mike waters and fertilizes the plant. The vine begins to grow.

Soon, there are lots of
big green leaves.

This is important because
the leaves capture the
energy from the sun to make
the plant grow strong and healthy.

During the Summer yellow flowers bloom. Some of the flowers have a small ball, or ovary, at the bottom.

When honey bees fly from plant to plant, they pollinate the flowers.

Farmer Mike knows
these small yellow ovaries
will develop into pumpkins.

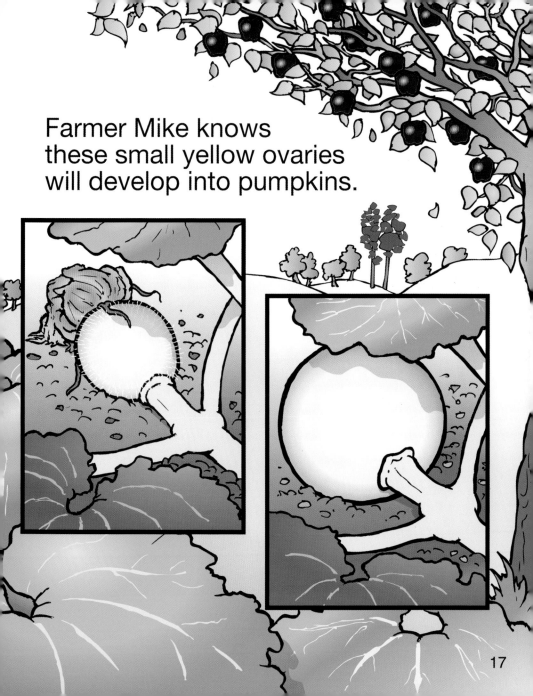

To get really big pumpkins, Farmer Mike cuts off all but one or two pumpkins per plant.

18

He also carefully moves the vines to make plenty of room for the pumpkin to grow.

He continues to water and fertilize the plant.

By early Autumn, the leaves begin to die back and the pumpkin has turned orange. **IT IS HUGE!!**

And what do you think he will do with the giant pumpkin?

Farmer Mike
carves it
into a big
smiling face!!

He also saves some
seeds for next year!

23

Farmer Mike Grows Giant Pumpkins

For additional information about
giant pumpkins, carvings, book availability,
or the Farmer Mike Awareness™ series,
check out our web site at:

www.QuarterCircleV.com